CREATURE FEATURES: CLASSIFY ANIMALS!

I0605268

IT'S A BIRD!

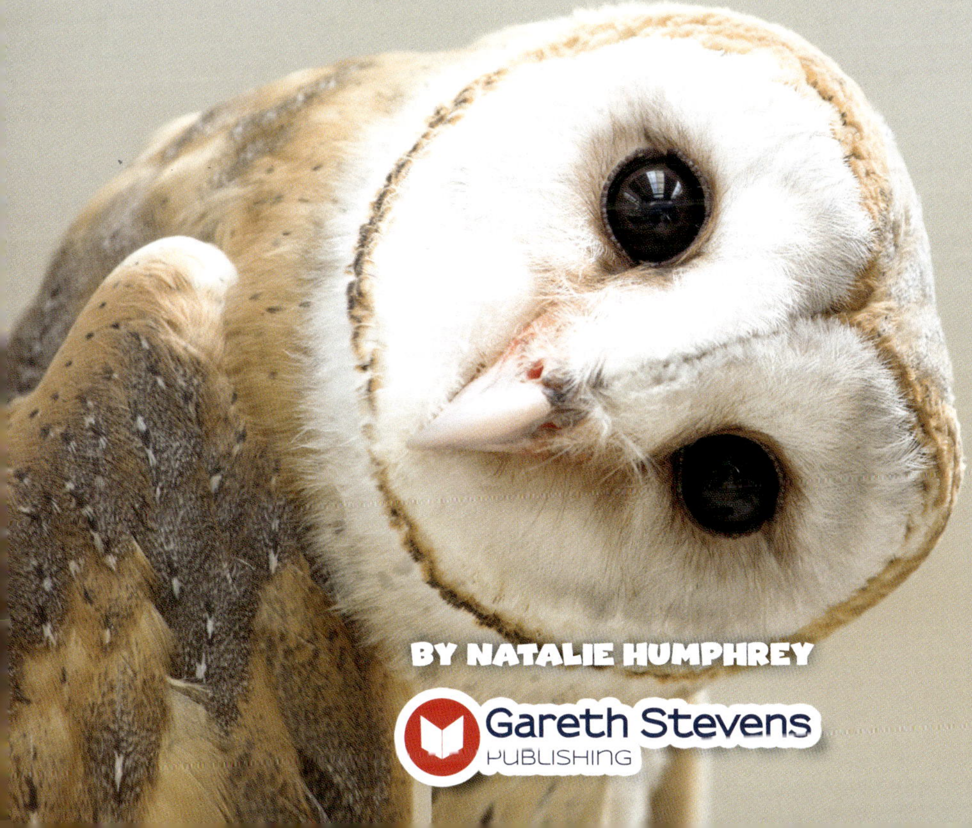

BY NATALIE HUMPHREY

Please visit our website, www.garethstevens.com. For a free color catalog of all our high-quality books, call toll free 1-800-542-2595 or fax 1-877-542-2596.

Cataloging-in-Publication Data
Names: Humphrey, Natalie.
Title: It's a bird! / Natalie Humphrey.
Description: Buffalo, NY : Gareth Stevens Publishing, 2025. | Series: Creature features: classify animals! | Includes glossary and index.
Identifiers: ISBN 9781482466805 (pbk.) | ISBN 9781482466812 (library bound) | ISBN 9781482466829 (ebook)
Subjects: LCSH: Birds–Juvenile literature.
Classification: LCC QL676.2 H848 2025 | DDC 598–dc23

Published in 2025 by
Gareth Stevens Publishing
2544 Clinton Street
Buffalo, NY 14224

Copyright © 2025 Gareth Stevens Publishing

Designer: Andrea Davison-Bartolotta
Editor: Natalie Humphrey

Photo credits: Cover Anan Kaewkhammul/Shutterstock.com; series art (backgrounds) Honyojima/Shutterstock.com; p. 5 SARATSTOCK/Shutterstock.com; p. 7 Evelyn D. Harrison/Shutterstock.com; p. 9 anitapol/Shutterstock.com; p. 11 (top left) Gallinago_media/Shutterstock.com; p. 11 (top right) CSNafzger/Shutterstock.com; p. 11 (bottom right) Toeizuza Thailand/Shutterstock.com; p. 11 (bottom left) Pulak Sarker/Shutterstock.com; p. 13 leopictures/Shutterstock.com; p. 15 (bottom) Ramona Edwards/Shutterstock.com; p. 15 (top) showcake/Shutterstock.com; p. 17 (bottom) Vaclav Sebek/Shutterstock.com; p. 17 (top) IHERPHOTO2/Shutterstock.com; p. 19 (top left) Karelian/Shutterstock.com; p. 19 (top right) Robert Linehan/Shutterstock.com; p. 19 (bottom left) Bermek/Shutterstock.com; p. 19 (bottom right) Wirestock Creators/Shutterstock.com; p. 21 Bonnie Taylor Barry/Shutterstock.com.

All rights reserved. No part of this book may be reproduced in any form without permission in writing from the publisher, except by a reviewer.

Printed in the United States of America

CPSIA compliance information: Batch #CS25GS: For further information contact Gareth Stevens, New York, New York at 1-800-542-2595.

CONTENTS

Is That a Bird? 4
What Do They Look Like? 6
All About Beaks 8
Bird Feet 10
Nesting 12
Baby Birds 14
Around the World 16
Living Near People 18
Helping Wild Birds 20
Glossary 22
For More Information 23
Index 24

Boldface words appear in the glossary.

Is That a Bird?

What's that flying high in the sky or hopping on the ground? Does it have a body covered in feathers and two **scaly** feet? If so, you might be looking at a bird! Birds are **warm-blooded** animals that can be found around the world.

What Do They Look Like?

Birds come in many shapes and sizes, but they have a few things in common. All birds have a **beak**. They have eyes on the sides of their head. Birds have wings covered with feathers. Most birds use their wings to fly.

WINGS
BEAK
FEET

All About Beaks

Birds that eat seeds and bugs have shorter beaks. Birds that eat fish have longer beaks. Birds that eat meat have sharp, **hooked** beaks. Birds with long, skinny beaks sometimes eat **nectar** from plants!

DIFFERENT BIRD BEAKS

Bird Feet

Most bird feet don't have feathers. They have a claw on each of the toes. Birds use their claws to hold onto branches—and their food! Some birds may use their claws to build a nest too.

Nesting

Birds start their lives in eggs. Most birds build nests to keep the eggs safe. These nests can be made of sticks, feathers, or even the fur of other animals! A mother or father bird sits on the eggs to keep them warm.

Baby Birds

After a few weeks or months, the baby bird is ready to **hatch**. Baby birds use a sharp bump on their beaks called an egg tooth to break out of their egg. Most baby birds stay with their parents after they hatch.

EGG
TOOTH

Around the World

Birds can be found all around the world. Some birds, like king penguins and snowy owls, like colder weather. Other birds, like parrots and toucans, like hotter weather. Some birds will even spend most of their lives in the water instead of on land!

Living Near People

Many birds live in the wild, but some birds live around people. Birds like seagulls, American robins, rock pigeons, sparrows, and black-capped chickadees can be found in cities. You may even find them in your own backyard!

Helping Wild Birds

While you should always leave birds alone in the wild, you can still help care for them! Filling a garden with **native** plants, turning off unneeded outdoor lights, and learning about the birds living near you can all help birds lead healthy lives.

GLOSSARY

beak: A part of the mouth that sticks out on some animals and is used to tear food.

hatch: To break open or come out of.

hooked: Curved or bent.

native: Describing a plant or animal that naturally lives in a place.

nectar: A sweet liquid made by flowering plants.

scaly: Covered in dry, hard plates or skin.

warm-blooded: Able to keep the body at a steady temperature no matter what the outside temperature is.

FOR MORE INFORMATION

BOOKS

Feldstein, Stephanie. *Save Birds.* Ann Arbor, MI: Cherry Lake Publishing, 2024.

Rzezak, Joanna. *1001 Birds.* New York, NY: Thames & Hudson, 2023.

WEBSITES

National Geographic Kids: Birds
kids.nationalgeographic.com/animals/birds
Discover different kinds of birds and how they live in the wild.

San Diego Zoo Wildlife Explorers: Big, Beautiful Birds
www.sdzwildlifeexplorers.org/stories/big-beautiful-birds
Learn about all kinds of birds, like secretary birds or the flightless emu!

Publisher's note to educators and parents: Our editors have carefully reviewed these websites to ensure that they are suitable for students. Many websites change frequently, however, and we cannot guarantee that a site's future contents will continue to meet our high standards of quality and educational value. Be advised that students should be closely supervised whenever they access the internet.

INDEX

American robin, 18
baby, 14
beak, 6, 8, 14
black-capped chickadee, 18
claws, 10
egg, 12, 14
egg tooth, 14
feathers, 4, 6, 10
feet, 4, 10
food, 8, 10
hatching, 14
king penguin, 16
nest, 10, 12
parrot, 16
rock pigeons, 18
seagull, 18
snowy owl, 16
sparrow, 18
toucan, 16
wings, 6
warm-blooded, 4
where found, 16, 18